SELECTED

SELECTED POEMS

David Harsent

Happy Birthday

18 / 4 / 90

All my love

Oxford   New York

OXFORD UNIVERSITY PRESS

1989

Oxford University Press, Walton Street, Oxford OX2 6DP

Oxford New York Toronto
Delhi Bombay Calcutta Madras Karachi
Petaling Jaya Singapore Hong Kong Tokyo
Nairobi Dar es Salaam Cape Town
Melbourne Auckland

and associated companies in
Berlin Ibadan

Oxford is a trade mark of Oxford University Press

© David Harsent 1969, 1973, 1977, 1984, and 1989

Selected Poems first published as an Oxford
University Press paperback 1989

British Library Cataloguing in Publication Data
Harsent, David, 1942–
Selected poems. — (Oxford poets).
I. Title
821'.914
ISBN 0–19–282602–6

Library of Congress Cataloging in Publication Data
Harsent, David, 1942–
[Poems. Selections]
Selected poems/David Harsent.
p. cm.
I. Title.
821'.914—dc19   PR6058.A6948A6 1989
ISBN 0–19–282602–6 (pbk.)

Set by Wyvern Typesetting Ltd.
Printed in Great Britain by
J. W. Arrowsmith Ltd., Bristol

To Roger Garfitt

ACKNOWLEDGEMENTS

Sections of 'The Windhound' first appeared in the *Observer*. 'The Analysand' was included in *Poems for Shakespeare—10*. 'Playback' was produced as a Greenpeace poster-poem, with an illustration by Ralph Steadman.

CONTENTS

A VIOLENT COUNTRY

Legendry

It was a violent country:
explosions and raw sunsets,
inexplicable cries,
glistening scraps
of butchered animals
caught in the forks of trees.

Soon the paths faded
into ferns; thornbranches struck
at our helmeted heads,
our silk and ermine
sadly tattered. It was
a quest of sorts.

Deep in the wood
we encountered the mythical child;
he was not pink
and smiling as we had been told.
He smelled of burning.
He was as green as ice.

We left him there—a bad
omen, like the sunset.

Simeon Stylites

for Michael Denning

Up here is a stillness I like.
At night my blood
cools as the sand cools,

by day the faithful ascend
to take my cracked
palms on their heads.

How my slow
death engrosses them!
They glimpse the glint

of sandgrains on the sere
plane of my cheek,
torque of my outstretched arms

settling to bless,
robes like a candle's
winding sheet

falling from the hard flame
of my head.
It is true and not true;

the pillar awes them,
and the slow climb they must make
into the raw

flexing light of the sun
to touch my foot, or bear
witness to my wounds,

makes them penitents,
makes me their saint.
They rise

and cull my prophecies
or beg for health,
then recede like inessential thoughts.

The high white wisps
of cirrus thin,
the sky

dims to indigo;
or there is a swash
of rain, rinsing the pelt

of sand out of my hair.
Nothing changes; and I
am glad of it.

Legs locked rigid, I
sit in my own stench,
my mind out on a limb.

I watch the writhe
of viscid maggots on the ribbed
tusks of my feet;

flies roam the collar's macula.
Far off, a hawk quivers
still, then drops;

the image fades, but not
the thought of death:
slow fall

of the sunbaked flesh,
the body's rot,
simplicity of the eroded bone.

Dawn Walk

Flints chink underfoot.
Worthless lode,
they litter the topsoil.

There is nothing here
for the birds
clattering in the hedgerow.

Their bald eyes swivelling,
they riffle their feathers
in the sallow light.

Spring. The whole earth heaving.
All winter long,
wrapped in their cauls of web,

their curious mosses,
the dead
have been practising

a whisper.
Ignore them.
They are the earth's junk.

Buds needle the thin branches.

Home Movies

The children raid the foreshore,
hunting crabs
and sprinting after gulls.

Standing at the sea's
edge, I watch you place
your bare feet among the pebbles.

You ignore my wave and stand
confused, trying to find
a painless path to me.

2

Stretched on the towel,
you are smiling.
Asleep, perhaps. Your skin
begins to burn.

I'm out of this.
I sit watching the placid tide
inch across a line
of pure white rock.

3

The spectral nimbus
round the sun
is a perfect, glowing circle.

Its menace
is clear enough to you:
you will not

look at it, or speak of it,
but gaze beyond the limp
curl of breakers, where

a sea-bird dives
for fish. Frowning,
you fix your eyes on this.

4

In the grey
mist of rain, the sea
is very still;
it makes no noise at all.

You kneel to gather
stones—those with
some sculptural weight.
Their blue

bulk appeases you.
Washing them free of sand,
you roll
their convexities into

the hollow of your hands;
your fingertips
amble along rough crevices;
your slow smile begins.

The Woman and the Animals

The same window.
That same window again.
She leans right out,

speaking softly.
The sound of rain
woke her,

as if she had
been waiting for this.
She smiles and leans

right out
to hear better the hiss
of rain,

pleased by the cold
drops she takes
on her up-

turned palms,
liking the way
it beads in her hair.

The fields are
black and still.
She knows the grass

conceals
the movements of
small animals;

it is to them she speaks:
softly, without desire,
without regret or grief.

She thinks
of them—their small
round heads, their sodden fur.

The rain quickens.
The fields soak.
She leans right out.

The Woman's Soliloquies

I

The low flames tire, contract to thin
loops of smoke.
The words she burns
ball in her throat and hiss. Reaching
to touch the ash,
she lets the crisp, black fragments smear;
the small contact appals.

'The children' she says,
speaking as if they were hers.

Their hot flesh cracks and falls.

2

Why should she be here, unless
it is to watch these gulls—
flown inland—
gliding and banking above the blank
length of a field which is filmed with white
like a dead eye?

'When I am old,' she whispers, 'the same lie . . .'

The gulls are utterly free;
they drift back on the freezing wind
but will not settle or cry.

3

She rises early to walk about the house
alone, astounded by the metamorphoses
forced by the silence and the tinny light.

Her small possessions, everything she owned,
she now denies,
liking the sense of loss.
Her image disturbs the glass.
Her soft speech stupefies the room:

'These common objects; all this dross.'

4

She is lying again. She knows it.
She likes the lie and listens to it
as if the voice belonged to someone else.
Her fingers pulse
on the globe's face as she sets it
turning slowly on its axis.
She sits on as the room darkens,
naming, one by one, the spinning cities:

'Paris, Rome, Vienna, Athens . . .'

5

'My fingertips connect
precisely, there;
they form a shallow cusp—
something I can bear
to watch, hour after hour;
it symbolizes silence—
prismatic glass of rich
anaesthetic colours which
refracts the feral light
in quiet beams of green
and holy blue,
the colours of dreamless sleep.
On other nights
there will be something else to do.'

6

'The season turns. The ewe
bleats all night, at dawn dropping
her lamb, bloody, onto
the dwindling snow.
The small changes begin
with subtle rain, with thin
vicious shoots.

            Tonight I lack nothing.
It is enough to sit in silence,
my whole body becoming
delicate and flawless

like a fine
porcelain Madonna, breasts
stiffening to shell and veined with blue,
my womb fragile and clear as glass,
sealing its frozen ovaries.
The vigil sanctifies—
avulsion of all that was gross.'

7

'In this house
of broken locks
and shattered mirrors, it is best
to sit still. My breath
does not disturb the dust.
My hands—that taut
arch where fingers press—
are neat,
clean, empty, passionless.'

8

Waking, again, at dawn, she hears
her voice fading into
the room's silence,
although it is hours
since she spoke. These dawns appear
as part of some immense
pattern, too vast for her to know.

Fragments of dream adhere,
like seaweed on
a washed-up corpse;
words she said, and words
she dreamed she said, obsess
her all day long.

That evening, tired of speech,
she tests
the astringency of song.

AFTER DARK

Two Postscripts to my Father

As you approached the farmhouse, came
the roar and sudden heat
and the moment when everything paused as you fell,
acres of sky
revolving and sliding away;
then you struck.
The world, you said,
shrank to a splash of blood as the pain began.

Your young wife, the child you'd never seen,
that cold house on the moor's edge:
how could you begin
to piece it all together . . . On the way
back to the field ambulance, you cried
feeling the steel
slivers bite inside your wounded head.

2

For twenty-five years
you kept those secrets. Dragged yourself
home from icy scaffoldings and plunged
pinched hands into the hollow of the grate,
or stumbled through the leavings of a dream
at three or four or five o'clock,
to thrust your head beneath
a gushing tap, wanting to soak
the pain away. And all the time
love beat like some mad bird against
the bars beneath your skull.

Now it is late October and the high winds have begun—
streaming through the stand of elms beyond my gate,
putting entire hillsides on the move.
Those early photographs litter my desk.
Smiling, sunburned, young, you lounge
against a thirty-pounder
or, hands on hips, stare straight
out at me. There's no way back
through all that violence. We live
seven miles apart and seldom meet.

After Dark

Strange that I should think of them so much,
those journeys he undertook: all those renewals.

I have bogged down in this odd village;
my children belong to the place.
At night I can hear the cows
coughing in a field behind the house.

I close my eyes
and invent arrivals.

Myth

His first clear memory:
the sky dull red at midnight,
a semi-circle pulsing into black,
and massive winds
roaring across the marshland.

The house was proof
against those far disasters.
Each night, in his mother's bed,
he watched the shadows looming on the wall,
a slow procession of defeated heroes.

Homecoming

A sense of history
unadorned, like algebra,
the main road empty at noon,
a clock
ticking by the fragrant stairwell.
And I am here,
a tourist in other people's lives.

My childhood
must be somebody's dark secret.

Encounter

It was a bad day for fishing; it really was;
but the sun flashed tinsel on the wave tops
and the breeze offshore was flattering.
Then, just before lunch,
the power boats rounded the point—
in the foremost prow, the image of his mother.
'Impossible' he muttered, but there she was,
lipstick, eyebrows, hair, her flowered frock,
just like in all the family snapshots.

Cockade

The silence of auditoria, the groan
of ropes on a midnight jetty.

In the stalls, a glove, or a black cockade;
triplets of backwash ticking against the boats.

These are the missing moments in our lives:
perfect, unenjoyed.

Here

It is grey rain
and you moving about these cold rooms
like an exile.

Morning comes
with a flurry of wings
outside the window. Wake up;

the valley is coddled in mist
as before. Wake up,
you can begin your vigil,

waiting for the dim
stars to reappear,
for the wind to sharpen,

for the last
slow flocks of birds above
this ragged skyline.

Out

The light craft had been buzzing past all day,
low in the water and built for speed.
From the crown of the fourteenth bridge
the views were endless.
Blue infinities where everything gave way.

She guessed there were risks in small indulgences,
even so, she was glad to be out.
The water pouring back beneath her feet,
the child beginning to move,
everything at blood heat.

Zoological Museum

The snow fox is snarling behind glass,
gums smoothly pink, fangs arched.

Above my head, an osprey is strung out

in full flight. Its wing-tips blur
to bright reflections of the neon strip.

Last night, you were somewhere between

me and the dulled horizon. I drove through
flickerings of moths, thinking how we

have spent our lives

these eight years. You'd waited up:
not drinking, but the room was foul

with cigarette smoke; and you slept

wrapped in a coat, your face, amid the fur,
fragile, as if you wouldn't last the night.

Leap Off the City Skyline

The first step
brought his face up to the sun.

Turning, he fell
through a hall of mirrors

where his eyelids silvered over
and the glow

ran on his face
like highlights on the sea.

That moment gave his purest breath,
sealing his lungs.

That moment
was his last glimpse of the sky:

a fractured blue, and clouds
like dark pools where his other lives submerged.

DREAMS OF THE DEAD

Truce

The dust in corners
sheds the scent of almonds.

The window is a mirage
where the dull
planets revolve around their own reflections.

The pattern grows
simpler with study, as she squanders
the last hour before daybreak.

Somewhere, beyond the steel perimeter,
her plane lifts off.

*

Memories of the flags,
the flea-market, the music.
Five islands where she lived
a peasant's life for most of one summer.

The icon has a smell I can't detect.
The colours seem to fade as they are watched,
fogged by the tinted lamp, blue becoming grey;
but still the child's nimbus
glows in the pale light.

*

Our first drinks
pledge a kind of freedom.
In the lighted windows of courtyard flats,
couples are raising glasses.

Seated on a cushion, glass in hand,
she watches how her Japanese
flower arrangement droops
towards the heat.

Those famished pinks
are all she has to learn by:
love's only accidents.

*

With dusk, a few stray snowflakes
taking hold
on windowsills and basement steps.

She wipes a bowl with garlic, lights the stove
and lays ingredients
along a glass-topped table.

The main line stations
are full of new arrivals
buttoning coats, and staring at the weather.

Five minutes, maybe less, and they are gone.
Their footprints on the pavement
gather a brittle lace along the rim.

The house grows quiet around her as she works.
The kitchen water
comes straight from the Arctic ice.

*

The lights of the city come on,
blue and orange.
Silhouetting nothing,
they burn like flowers that only bloom at dusk.

Whenever she is ready, they are there,
framed by the window.

The sea; the place of her birth; the lights of ships.

*

Perhaps it was
an act of courage simply being there,
knee-deep in the blinding white.

The photograph
shows her flanked by trees filled to the brim
with snow, like goblets;

she is gazing
away to the left, composed against
a perfect backdrop.

Her cigarette,
dabbed against the fire's filament,
browns along its length.

'That winter
was the last of things,' she says
and frowns, and feels ashamed.

*

The two sisters across the street
spend all day in the kitchen.
Their dark, identical shawls
smudge the window as they watch the stove.

Below their balcony
an awning flaps like a sail.
A man in dove-grey linen gloves
is sweeping sodden leaves over the kerb.

Sunday afternoon.
Her future falls into place.
She smiles at the thought, then comes awake
for the aftertaste of limes along her tongue.

*

Denim and pearls; her nails
left untouched;
her eyes in the ancient style.

The one blank wall
is freckled with morning sunlight:
maps of last night's promises,

a circus in Isfahan, a pine forest,
the drive cross-country
to avoid the rains.

Her palm-print
shrinks on the mirror as she turns away.
The day is spoken for.

*

Closed shutters; doorsteps growing grass;
the twilit stillness of deserted hallways.

Attics strewn with the broken ghosts of sheeting,
with family albums, relics of a war,

and toys preserved intact by gifted children:
the quiet, abandoned things of settled lives.

Rooms she imagines. Worlds beyond her own.

Disturbed by silences, she lies awake
trying to recall what fixed her here.

A lullaby, perhaps, the smell of uncut lupins,
a bus ride, a refusal, lines of spray
along a sea-wall, milk in brown stone jugs.

*

She comes in past the heavy gilded mirror
smiling, wearing a wide blue hat.
I rise and start to speak. The door closes.

She enters the room again, smiling,
wearing a foxfur and the wide blue hat.
When I speak
my words fade on the air; the door glides shut.

She comes into the room,
smiling, wearing the hat, the foxfur,
and I rise.
                    —My dream about the room.

Now she comes in, slowly, balancing
honey in a dish and melon slices.

*

Dog-tired, but somehow keeping track,
she sets up small objections to the music
but lets it play.

Crouched by the fire
she soon ceases to listen;
her eyes are watching something miles away.

If nothing's changed
an hour from now, we've won:
survivors of the wind, the streaming glass,
the life outside.

Endurance

The orchard is lit
on one side by the evening sun.
Its yellows glow.
An odour of leaves and rind pours in.

Hillsides are dealing with last year's scars—
endurance of the just-visible.
Twin children, hair ablaze,
number the lunatic eyes of moon daisies.

It is the true magic.
In the pause between mind and movement
history claims its own.

Dog Days

Summer's high-spots are burned out.
The dry slopes facing south
bear trenches of black scorched into the stubble.

The boys watches his nurse
as she confiscates the flowers. Her hands are stained
just above the wrist. She smiles her smile.

The sheets warm, the sun on the windows opposite
blurring at the corners of his eyes,
he feels his senses losing ground.

The house filling with murmurs.
A drift of woodsmoke past the glass.
The *chink* of bottles brought from the cellar's coolness.

Fishbowl

for Fred Taylor

I

A couple are stepping out on platform four;
his campaign ribbons gleam, her breasts
roll beneath the satin. Nests
of tourists spread their maps out on the floor.
The patterns alter. There's a law
that governs departures, and one that governs murders.
In the gloom of the roof workmen are welding girders.

2

A 'northern airstream' puckers the parkside trees.
The animals in the zoo lean on their bars
or pace in heady circles. No-one sees.
The rich arrive, downy in perfect furs,
at theatres with neon-etched marquees.
The wind that killed the Brontës. Iron spears
ring the wolf enclosure. There are pleas
for one more encore from the velvet tiers.

3

The bald man mutters across the rim of his glass
to the girl in the stetson; she has brought her dog.
They're nothing special. Except for the sunny gloss
on his skull, they'd be lost in the multicoloured clog
of drinkers and talkers. The neatly bevelled fosse,
that drains into the river, froths with ale
and scraps of muddied bunting from the farce
(the players have left, taking their cash-on-the-nail).
The party revolves like a slowly sinking wreck.
'Perfection . . .' he whispers, resting a hand on her neck.

4

Couple by couple, they find their place in the dark.
The girls know how to glide beneath a touch:
sleek and unevolving, like the shark.
The piano-player downstairs knows too much
about the gritty business of the skin;
even so, he likes to leave his mark.
Angling her knees, she reaches to guide him in.
He yelps as his spreading fingers brush her fin.

5

Sheep nudge and nibble the pallid downs. The hare
circles, in a frenzy, back to where
women in bowlers and breeches are crossing a stream,
flexing their thighs as they hear the quarry scream.
The bank is pocked with the marks of paws and studs
that will garner scabs of ice from the freezing air.
The season is right for worshipping cruel gods.
On a mattress of coats, acting a part from a dream,
Wednesday's child is screwing his girl in the woods.

6

She smiles and smiles at his raw, grogblossomed face.
Smiling makes the good times come.
Dance-floor partners stall and change the pace,
their stomachs touching and their heads in space.
His dampened finger tamps a crumb:
he licks it, gives his empty plate a shove,
and looks up smiling. All we know of love
is pain and the response to pain.
They waltz in the old-fashioned way, with impulsive grace;
he leans back to speak, and she smiles again.

7

Laughter in the Wendy house, a ghost
of someone's face reflected in the pane.
Love provides the terrors that we hide,
then seek . . . the party's nerveless host.
The scents that gather in the gathering rain
bleed off from the garden like a tide,
spilling across the pavement. Parents ride
in convoy to the house; insane
visions make them shudder as they coast
along the driveway: lights through a seepage of fog.
The children hoard their secrets, safe and smug.

8

They are slender-waisted, blonde, with eyes like glass.
Lodged in the grass, they watch for rival styles.
She smiles as he gets up on his knees to piss.
Crook a finger at bliss and music storms
between the tents where, adorned in pretty rags,
she sighs and sags against his bony chest,
rumpling a nest amid their tangled rugs.
Sometimes he begs; she sometimes flirts with theft.
His pinkie browses the cleft between her legs.

Dreams of the Dead

April 30

The women started downhill from the crest
sidling against the slope, their skirts
lifted at the hip,
their braided hair
releasing wisps of light into the wind.

The morning opened out with blues and golds,
the smell of chlorophyll.
                              The sun
strung opal beads along
his eyelash as he woke.

*

May 2

Hottest near the surface: vast white arcs in space
the sunspots crack in blackness.
Tidy in bed, his arms and legs aligned,
he recalled how swifts will sleep while on the wing:
a night-long, dreamless glide.

The room grew warmer. Dust rose in the sunlight.
Opposing mirrors, blank
and fathomless as water,
waited for his image while he slept.

*

May 7

Jungfrau. Snow slopes near the peak.
Each snapshot slightly fogged
by cloud and background movement.
Inside the Ice Palace
light struck in frosty lines across his eyes,
the silence like a hood.

Thousands of feet below
thin streams lace the lower slopes.
Sure-footed on the scree, the women move
down towards the foothills, the altitude of songbirds.

*

May 9

Like spiders' webs unpicking,
something crackled deep inside his lungs.
Those mornings when his hands shook
and that same
bitterness rose in his mouth,
he let the dreams fade as he roamed the house.
The slow roll of a cypress in his garden,
or a crease
of rain across the glass, would be enough
to turn his thoughts.
                    Then coffee and the papers:
the fire in Asia, the small world of statistics.

*

May 15

Mornings arrived with tiny gaps in meaning.
The spring winds dropped at dawn:
time for the garden's old inhabitants
drifting between
the glistening boles of pear trees.
Remnants of the first rain jostled down
from branch to leaf, splitting the early sunlight.

'I live for those moments when we both
forget ourselves.'
                    He dreamed he woke,
seeing first the waxy curlicue
of a lily under glass,
a pomander,
a painted eggshell resting by the mirror.

For a moment he held them in focus.
Old bargains. The death wish in women.
When, finally, he surfaced from the dream
the streets were dry, the morning almost done with.

*

May 21

Albino. Poor child. He tiptoes through playtime,
his bloodless head
in negative against the wood's penumbra.

Where the sun strikes back
from the blazing cruciform of the chalk excavations,
he almost disappears.

His constant companion, that dark boy,
crosses the wiry turf
at a dead run, his brown hair smudged with leaf-dust.

The women smile, shielding their eyes to watch.
They are laying out
fruit and wine and bread on a damask cloth.

*

May 22

One mirror is sealed over by the sun,
a brittle barrier
where something is trembling for release.

A flight of doves, lit on their undersides,
wheel into view, then clatter overhead.
Too tired for that, he lets his eyes slide shut.
Fragments of retina
swim up beneath the lids.

*

May 31

Five massive rocks
chopped the undertow to ragged
bursts against the blue;
that vicious seething was the only sound.

On the farthest lip of sand,
a flutter of colour, the women's parasols
spread pools of shade. Softer than breath,
a pulse ticked in his temple as he watched

then turned and went full-tilt
towards the breakers. Engulfment in that green
drifting light took hours: the throb of water,
the mild tug of the tide rolling him on.

*

June 2

His senses trapped,
he watched the views float back,
sharp-edged and colour-perfect.

Most of one night, sleeping and awake,
rehearsing someone else's tragedy.
The late train to the coast,
the mourners grouped
on one side of the treeless, cliff-top graveyard;
and then, at dusk, the house awash with light,
a figure on the terrace
standing still, listening for the rhythm of the sea.

That sound set him afloat, his body caught
the motion of the swell.
A figure on the terrace,
looking out.
The salt taste of the child's lips.

*

June 7

Above the slender elms, the local-stone
period chimneystacks,
the spire cut a wedge into the sky.

Lightning always struck there first.

A man had died one night,
drunk and desperate,
roped to its perfect profile. The churchyard crows
scored his cheeks next day
before the thatching ladders could be brought.

A peal of six bells, echoes into echoes.

Was that a dream as well?—
The Virgin's mantle in the eastern window
staining the steepled hands;
the women's veils
stirring as they mouthed the first responses.

*

June 9

Early-morning cloud banks nudged the hills,
the underside of heaven,
misting the true vanishing point,
the point of departure.
                      The blues and blacks in nature
absorbed him: those shadows, the stiffening wind,
and then the first thin stalks of light.

Rays to redden women's hair, a breeze to make it fly!
Unimaginable, the solar winds
roared through space, putting the earth awry.

*

June 10

For seconds at a time, his mind stopped dead.
Not faints, not *petit mal*; a loss of will
perhaps. He shuddered, then emerged.

Three nights without sleep had left him dull.
It seemed his friends, his generation, dwindled; some
had made a case for love—

the age would not permit it.
                              'But it's you
I fear for most. What's to become
of your life now?'
                    A memory of the sea,

the southern rip-tides,
filled his head with noise.
His night-light dipped and flared behind the glass.

                              *

June 12

Eyes bright with fatigue, he walked to where
trees scattered the glare, a roil of leaves
flowing along the bough.

Six horses cropped the downland.
A gleam like silk
travelled their flanks. He dozed and watched.

In single file
the women left the treeline,
a flicker at the corner of his eye.

He blinked. They passed between
a stone wall and the wheatfields
into the shade, the blue bowl of the valley.

                              *

June 20

The children's voices teased him,
always just beyond
the next dune, or the next warp of the cliff.
How long ago was that? They learned
to walk lopsided on the pebbles
towards the shimmer where the sea fell back,
a mist of blue and amber.

He dreamed the place again:
in moonlight, calm, deserted, with a sea
flat, black and frangible.
And all at once
was certain of a desperate energy
beneath that placid surface—
of a form,
teeth clenched against the salt,
swimming upwards, wildly, through the dark.

*

June 23

The summer lasted, month to month.
Neighbourhood children tanned
and ran the streets till evening.
His household glass
beamed chips of rainbow on the walls and ceilings.

All brightness taxed him;
the blinding linen
that lapped his face each morning,
the glance that scorched his eye,
the inexhaustible
crystal at the centre of the sun.

*

June 25

He broke
eggs into a pan and watched them settle,
perfectly smooth, a fleck
of brighter red in one; energy arrested.

41

(The sounds of five a.m.:
birdsong out of blackness,
a wind riffling the orchard).

The dreams that spoiled his sleep
still kept their hold:
the kitchen—warm with cooking—
hung everywhere with women's implements.

*

July 3

The sweep of their skirts leaves tracks in the morning wet.

One whistling a tune, and one
hanging back from the rest
to watch a poppy's petals smudge to mauve
as they are torn . . .

Brightness like a backwash draws them on.

Bones into chalk. Pelts into melting pap.
Victims of the twilight kills
treading to mulch and fragments underfoot . . .

They harbour a gift: decades of memories.

Their high-necked blouses dampen,
a pool
of moisture at the hollow of the throat;
the cling of cotton at their calves and thighs.

The wakening town hangs in a pale haze
beyond the valley's rim;
they keep it always in sight.
A pair of swifts are shuttling across the blue.

*

July 9

An ensign in his dreams,
the storm beacon
flew each day above the tallest outcrop.
Each day
the sky's first flush turned indigo by noon
and a wind poured in from the sea.
                               The house
stood four-square to the weather, windows flexed,
the verandah's raddled stanchions
soaking up stain.
                  That much was familiar:
light teeming across the bay,
a building rooted in stone,
a figure on the terrace, looking out.

Later, the focus altered—something new—
to sandflats furrowed by the ebb,
gulls and clouds and shallows,
inshore rocks,
all fixed, an instant, in the frame.
And barely visible in so much space,
that form, beached on the dunes,
shrouded by seagrass,
pungent with shells and brine.

*

July 15

He lay in the sun, motionless through the day,
a blizzard of red
roaring behind his eyes, while his skin
tightened and dried.

—The scorch that withers greenness, melts
the flesh from cattle, leaving them bones and leather,
that burns its starvelings,
bleached the orchard at its boundaries.

That night he shed a glow,
a child with a fever needing a woman's touch.
The solar gold
flooded his sleep, the whole life of the dream.

*

July 20

His sleeping eye
flickered to show the white; he saw
his own feet tamping grass, and pale moths
rising from the stalks.

He traced the lip of the hill by its darker line
on the sky's near black.
Miles beyond that ridge
the streets of the village were strung with coloured lamps

raising a glow, like beacons starting up
on peaks above the farmland.

The chalk slopes on the sheer side of the hill
were struck with the shadows of trees.

He waited for the sense of place to come
as the first flare cracked and rose

*

July 20

a sudden livid red
dipping towards the shore.

The cliff-edge slick with spray, the signal fires
flattened to orange circles by the wind,
acres of cumulus
streaming off the skyline . . .

Just to stand full-face watching the sea
numbed him with effort.
He knew the rest, a drama of repetitions:
death by drowning,

the drift and muffle of narcosis.
The body washed in,
a froth of limbs and spume across the rocks.
He screamed and seemed to wake.

*

July 23

Insects
ticked against the glass
or found the open fanlight,
trailing in
a flicker of tendril legs.
From somewhere in the house, the sound
of water trickling.

'If you die, I'll hear of it
and arrive to parade the churchyard
in my blue hat
pinned with flowers.'

The day assembled clear shapes in the heat,
outlines and edges;
colours sharpened through the early haze.

'What more
can love do to us now?'

*

July 24

The day gone grey with rain,
a thinner spray
blown like spindrift through the fall;
each room uninhabitable
as if the house had seen a death.

A keen eye could have picked them out
in moments when the downpour slackened,
crossing a high field beyond the town,
ranged like climbers.

He meandered back to bed,
wanting to sleep.
The whole eastern valley wall
glowed as beaded water caught the light—
luminous green when the cloud broke.

*

July 29

Tiny scavengers broke cover
from sour corners of the kitchen garden.

His room built heat
behind sealed windows; roof beams cracked and settled.

There were piles of books
stacked beside the bed. There were ashtrays,

soiled clothes, the sightless mirrors.
There was his small collection of knives.

'. . . you are talking of freedoms I haven't come to yet.
Nothing abides. But the days out here seem endless . . .'

*

August 6

They smile at the prodigal: women of the house.
Love and the warm brick and gulls on the lumpy fields;
nature feeding and the autumn dregs.

It dimmed and went awash,
as if his eye
had filled with blood or caught the flooding sunset
running red on hanging copper pans.

*

August 13

Blind and numb, wrapped in the caul of blankets.

He woke in fear,
seeing at once the widening pool of sky
smudged with small birds that twitched in ragged flocks
from pane to pane.
                          The evening before
his eye had followed the tracks in the half-grown crops
that seemed to flow back to a second, deep horizon.

Was that dreaming?
The daze of concentration, his slowing pulse,
the spire's knife-edge against the thinning blue . . .

*

August 23

The orchard beyond his window took the sun,
blurring the edge of everything.
It stunned him to look. The wash of distances.

In the spaces between the trees at the outer edge
the women waited, the light
lancing the colour from their eyes.

How strange, to love only the dead.
There was nothing to hear or touch;
there was nothing to be kept—

the room and its brutal ornaments,
his jumbled books,
the mirrors dissolving in each other's gaze.

He stepped into the vast, unshuttered glare
like a swimmer loosing his foothold,
moving through silence, his own element.

47

Bodinnick

'Love God and work the sea.'
His raven's voice
rasped from the pulpit.
Shadows spread in the chancel.

In the granite caverns
off Lantivet Bay
drowned men jostled stalactites;
their eyes, scoured by salt,

looked inland, where you set
your face against a gale
along the cliff-walk,
richness underfoot: wren's bones,

lardings of kelp,
the dusty, mouldering hides.
From the harbour, we watched you
dwindle on planes of blue.

Fishermen, like penitents,
trudged the shoreline
waiting on a tide,
the great shoals massing,

a sudden heave
of silver against the swell.
At twilight, diesel smoke
rose on the hill like ground-fog.

The child laughed in his sleep.
Hooded, you withdrew
into a flush of rain,
the cobbled squares around the quay,

palming a cigarette,
then returned with that dead look.
We lay above
the dull lights of the village

silent, burning with loss,
hearing the house beams creak,
the long tremor of rollers
ringing the corner stones.

Level 7

The lamps of the terminal cases burn till dawn.

*

The last of the storm
trails into the morning's undertow.
'Up here there are no horizons; we sleep among clouds
or watch the night sky, like renegades, for a signal.'

*

Ancient walls hold up under the weather, their granite roots
locked into the hillcrest; the city fathers are there
under chiselled slabs, serenely falling to pieces.
A century's corruption darkens the stonework.
Each evening, visitors' cars whine up that hill
kindling the crumpled windows, irrigating patterns of frost on iron.
Convalescents watch them hit the rise, headlight beams
toppling into the valley—then the whole sad convoy
cruising the outskirts ... love in the guise of fear,
hands closing on one another, cut flowers staining their tissue.

*

'... it was long ago, too long to get things straight;
in another part of the country ... and the people are gone.
So frail—I think my flesh could crumble,
flaking away from the bone like something cooked ...

I only remember
the moments of deceit ... that sudden shout of laughter
from the lawn outside my window; and later the girl
coming into my room, white as a fish,
smoking with powder after her morning bath ...'

*

The cloudless, endless blue begins to tear.
Instruments scald in their vats. The first drugs of the day are
 soaking through.
Stars make soft explosions beyond the dampened glass, their
 glimmer shrinks
back into space, a final rush of light
where the rim of the universe thrums like a wire in the wind.

*

Birds are thronging the undergrowth. Pale rods of sunlight
root among the groundsman's hothouse tulips.
In a shuttered room, an Italian orderly sings
as he washes the limbs of the new dead. Healers
with brain-cells in their fingertips
greet the day with knives . . . floodlights swamping the tile,
the blemished faces ebbing under gas.

At the Solstice

The lip of ice
yaws, enticing black runnels,
then the pane
splits with a sweet crack.

Frozen in. Your sinew
melted; but I imagine you
staring straight up
out of the cloak of hair

and subsoil litter,
fingers laced on your breast,
the impossibly long,
perfectly fluted nails.

We buried you
in the blizzard of '63.
It's an accident, coming back
in this feathery rush of snow,

the year's first ice-wind
strengthening with the dusk.
I needed a map
to find you. My youngest child

stamps tiny cleated footprints
along your neat verge
and examines the bright
curl of his breath on the air.

I've been afraid for so long,
laid-up in that place . . .
Nothing could cool me,
the sisters of steel

plied at my bedside,
gleaming and pleased to be there.
Each time I woke
with their voices hard at my head,

their talents were sprouting inside me.
As the needle drove in
and I counted backwards from ten
they would smile at each other, keepers

of some forgotten art.
The wickerwork cornucopia
spilled on my locker,
arranged as for Caravaggio;

the death tents
were pitched along the ward,
steamy with sickness. In each,
a crumpled pink bud, bedded down.

Drugged and reassured,
I can think of nothing to ask for.
These scattered flowers
are already brown and crusty,

little enough to add
to your deep dream of riches.
Light-footed on the frosty stone,
the Godly glide past

to Evensong: the few
friends who outlived you.
The pure note of the choir
calls them in,

the tapers, the purple and gold,
the gloss of decadence.
Like you, like them, I shall grow
each day more distant,

watching the future close down,
sensing malevolence
in weather, in rough dreams,
the old portents.

My voice will crack
the clear syllables;
my children will watch me, and let me
pretend to earn their love.

Moments in the Lifetime
of Milady

1

The island grasses were filled with unknowable scents
and the flurry of creatures running before her footfalls.
Even at dusk, the nub of the crumbling abbey
was a view on the eastern side.

On knees and palms, swaybacked like a stricken runner,
her skirt thrown up to billow across her back,
she begged, 'Pleasure me, pleasure me.'

He half-understood what she meant. As he worked
he looked out over the broad back of the island,
its lumps of ruin, its rockfalls to the sea.

2

They are gone, she thought, *my pale cavaliers are gone.*
Dogfights in the blue, the silken scarves,
the music after midnight.

How I miss them. Ice in the vein. A goblet's bowl
sang beneath her finger
as the room grew truly sensitive to sound.

She invoked a silence with cool dribbles of wine.
In its niche, her portrait drank light,
ambers and black, and green in the drape of her shawl.

3

Wading through clear water, her body foreshortened
to tits and tuft and her feet spread on the chalk.
The convoys had slowed them, rolling up to the border
under a veil of dust. The last of Europe's summer;
the scum of the earth dividing the earth between them.

The sleeper on the bank sweated his wine.
My paramour. His city had grown dull,
his voice, each morning, a clutter of vowels ... 'Tesoro ...'
She arched in a dive and swam for the riverbed,
a pale shape, sounding, sleek enough to be fished.

4 *Berlin, 1949*

Lovers hung like shades in the bar-room glass,
faceless in one another's grip.
She raised her drink to give them 'happy days'.
Offstage, the Joker winked a violet eye
and stripped amid the bottles and the smoke.

A sky of purple flock, the lights, the towers ...
Later that night, on the Bridge of Unity,
she watched the city die down;
unstrung by the brittle stares of the enemy,
the unopened letters in her hotel bedroom.

5

The barrels moved left to right, a perfect track,
and then the birds went ragged in the air,
bombing the black stubble.—Ultramarine,
September's washy sun on her back;
she moved amid the crossfire like a cat,

pulse jamming her wrist, a purple arc,
tributary to the flood
that funnelled through her womb. Gunshots
banged in her brow,
hosannas racked the trees. *The old
drool onto their pillows; this fierceness is lost them.*

6 *After Giacometti*

The deceived—they are the true obsessives.
That night, her smile
had finally wilted beneath the driveway lanterns.

She was always going to remember
the way the old couple and the children stood
between the whitewashed pillars

and how, as she moved, they dwindled to wire and wax
in the bolts of light
where their heads and hands were fixed.

54

It was spring, she noticed; the damp had a smell of the sea.
The gardener's sapling birches caught the night wind,
swooping shadows on black.

7

Her yellow roadster stood in the drumming rain,
a mist of white along its canvas top.
Inside the spinney a cool damp conjured smells,
blood-world of the weasel and the shrike
where she bunched her skirt and spread her knees in a squat.

The old life made her feverish:
a drone of voices from the lawn, rich wines,
father's chessboard set for play in the summerhouse;
the family bound to its ritual of sorrow.

Now she was clean away.
The day ebbed in her spine as she watched the stream
trickle between her feet into the thorns.
Smoke among the roots; meat on the spike.
A bludgeon came at the trees ... She could still taste
sugar from the child's kiss.

8

Rustling in the sill, the spider's ghost dance.
The place was for those who trust severity,
a whisper of salt on stone.

She brooded on her demesne,
mistress of the house, its oaks and velvets,
the garden's stems and saturated apples.

Her deathbed silks
slid from their tissue to drift across her palm.
What sin to hoard them! What peace to be beguiled!

9 *Madrid, Joselito*

Thinning pearl above an even skyline
where the storm was pulling back;
her train clattered towards it
through acres of orange groves and bamboo windbreaks.

That afternoon, half-choked by the heat,
she'd watched the gipsy killer take his bull,
gliding across the horns.

Christ, how I hate the south. Warm rain like floss,
the baking, purple dusks,
the women in their blacks, courting betrayal.
The train flushed dogs from the scrub. *I'll change my life.*

10 *K622*

Static drowned out the first few bars—
a feint; and then that clarity like sunlight.

The arm that looped her neck
flexed and relaxed. Watching as he dozed

she felt the music clutch . . .
Nothing would save them from the plans they made

at that first rendezvous.
Their deep, slow-burn to frenzy made her ill.

11

She licked the fontanelle . . . the boneless limbs
stirred and lifted like underwater stalks;
a thread of blue
tapped and tapped to irrigate an eyelid.

It was thus in dreams, it was thus
in all her best imaginings: the child
taking his strength in sleep,
his dampness on the muslin, the scent of milk.

Beyond the starched reflections in the window
the compound's orange glare
lapped the high fences. Madness on her lip.

12

They kept the room slammed shut against the heat.
Day-long twilight, her body a pale smudge,
spreadeagled, as he brought the cigarettes, the wine.

She watched him pour, then pause; some bleakness struck him;
he stared as if his face had emptied out.
Too late: the sudden, chill suspicion

of bad luck in the blood.
As he moved on her again, she softly spat
three times into his mouth, and sealed her passion.

13

Worn, worn to an edge, her eyes burned out,
salt in the reddened creases of her skin,
beyond words for it all, beyond weak dreams of peace
or deliverance, she drove at a tipsy crawl
to the house in Oxford, clipping kerbs,
dazed by the sunset puddling in the road.

All the old things—she toured the place like a stranger:
amazed; it was cold and intact;
her garden nudging the doorways,
dark ooze by the apple tree stippled with massive paw-prints.

'Wolves,' he laughed, 'wolves,' as they stretched on her
    childhood bed.
Later, he posed her by an open window,
'Keep that whole side tense. Keep looking down.'
—Into the garden, half-hoping to see
a ruff of hackles sloping past the gateway.

14

A three-day downpour, dissolving hills in mist.
After the abandoned conversations,
the sound of water like a long echo.

Sometimes, late at night,
she could almost abandon every wanton scheme . . .
dozy with drink, as the cats
arched and circled before the open fire.

'We are exiled here,' the young man heard her say.

MISTER PUNCH

Mr Punch

The tiercel feathers upwind
breasting the airstream;
there's game on the slopes.

Guests on the valley floor
spread their picnic cloths—
midgets warped by the mid-day haze.

In the tangle of brush near the pines
unseen animals go belly-down,
hot beneath their tawny pelts.

White linen on the furze,
ribbons and flags,
skyline clouds a belladonna blue.—

The stasis breaks. A child
is tapping a wineglass on his teeth.
His friends join hands to dance

as lust explodes in the bracken.
Mr Punch worms into the girl
and she squeals like a peccary.

Over his shoulder, she glimpses
the hunting bird
plumbing an acre of sky

while her blood heats.
Mr Punch is growling; the breeze
cools the sweat on his flanks.

His wife and family feast
and watch the dancers.
These women are all alike to Mr Punch.

He'd like to own them, he'd like to eat them whole,
he'd like their murders
feeding his night-time conscience.

This one's something special:
she loves a dare
and Punchinello thrives on secrecy.

Suddenly, horses are there on the hillside,
standing by their shadows to feed.
Distant, they seem to be

behind a wall of glass.
They peel off, circling the dark patches of clover.
A dozen faces look up

struck through with happiness.
White light flashes from silverware,
the wineglass sings

along its breaking-line.
Mr Punch emerges, grinning.
The children dance in their ring.

Punch and the Judy

He feels so old, something primordial,
something that surfaced through the permafrost
sliding blindly towards warmth . . .
Icy against her back: she dreams herself
diving through breakers in a winter sea.

Rain at three and rain again at seven,
hanging leaden in the tidy square.
Dawn after dawn—detritus from the whirlpool,
the spars and splinters of shipwreck.
Walls of water roar beside the windows.

The girl's blonde head is drawn
into a caul of weed
and her long legs trawl the dark.
His shoulders rap the sea-bed. There comes
a noise like singing as their bodies sunder.

Picked over by dabbing fish.—
Her plump lips on his face and on his neck,
dampness of hair uncoiling.
His mind comes loose: he sees a figure
out on the drowning streets,

camouflaged by morning twilight,
watching the room, his eyes
luminous, like an assassin's.
Her shadow runs on the curtain, then she floats,
a tangle of pink and gold on frosted glass.

Love is his energy and his trap, spurring
the thug beneath the skin: homunculus
hooknosed, hunchbacked . . . Her voice
rings in the shower . . . It stirs in its cage of ribs,
inarticulate and murderous and mad.

Punch and the Passing Fancy

They had the same name
for each other.
They evolved
points of recognition.

Marking distances
that narrowed—
in the street, in the park—
they gained on themselves.

Always the first
to wake, she'd sip
his breathings-out,
crusty and toxic;

he would soak
in her dew:
a vagrant sleeper
in some drenched place.

They unearthed
mysteries of kinship,
something quick
in the blood,

witching, venereal;
their hermitage
echoed, each day,
to cooing and crowing.

They traded
only in bliss,
taking their haul
to cafés, to theatre queues.

By firelight
she told stories
of the antedeluvian
world of friendships—

here were the sole survivors,
passionate mutants
too fragile
to prosper or breed.

Punch on the Boul' Mich

Girls in the Luxembourg Gardens, girls on the boulevard.
The appalling tyranny of unfoxed loveliness.

Punch relaxed with his anis. He conjured
a parade of vulvas, threshing moistly.

He picked her out by the way the air
enveloped her—a dog's recognition.

She mooched through the leaf-dapple. In her poured denim
and narrow boots, she was love's be-all.

Later, she almost fainted
on the precipice of Notre Dame. Punch saw

a body spread on the coping beneath the rose window,
still fashionable in death: the anglepoise wrists

and nerveless backward look
of shop-front dummies on the Rue du Bac.

The gargoyles clucked. They looked like relatives.
She swayed but hung on, swooning at verticals.

With a flick of his fingernail he stole a note
from Quasimodo's bell—old hunchback, brother,

innocent. The flow of the pitch
as it climbed made her eyes roll back.

In the hour *entre chien et loup* she came into her own.
Neon streaked her hair. The whores

turned from her line of approach, as scavengers
scent a predator. To amuse him, she walked ahead

canting her hip-bones to make her haunches roll
and he pictured her body's bedrock, the uterine siftings

a seepage between boulders. When she settled
at a table, he sauntered past

then backtracked to play the pickup, noting her face
as docile as a print on glass.

Punch and the White Madonna

She watched them torture each other
with money; her flawless eyes
detected loss before all else.
In the calms between curses and tears,
or during the torn-off kisses, she'd record it.

Spirit of the household,
she counted the drinks. One morning
there was blood on the floor by her shelf,
as if she had drawn it.

A face
perfectly set for enduring.
The cheekbone moulds
the cheek to glacial smoothness and her cowl
falls in haggard lines beside it.
Her lips have been scalded white
by his terrible rages.

Any thought of the flesh
dissolves before she can hold it.
She would overhear
the wrench of love-cries from a farther room
and read them as pain, but share
his dreams of the unmarked features of children.

She suffers the household accident.
A slip of the blade
will tender a sacrament.
He offered the bread, then ate
with the musty taste of water to seal it.

The faintest music
rolls in the brittle dome of her brow,
a whispering
gallery of woodwind sounds.
She would gather a note and hold it
so long that its passionate tremor
threatened to crack the marble, so long
that its constancy was a silence
like the silence that lodges between planets.

66

Sometimes he felt she might almost turn
to a tap at the window,
a clock-chime, or the sudden
sensation of shattered plates.
The movement lies
in her, but she can't perfect it.

Beneath the white headdress, her head
has the brow-band and chin-band to bind it.
The strapping confines her
as comfortably as a small illness.
Any gesture, any word,
would blemish her. She abides
in the mysteries of quietness, the quiet
gaze, the quiet, cruel
passivity women are born to.

If he drinks himself to sleep, she mulls
the patter of his dream-talk
while he flounders on the couch.
She wonders at it,
but neither flinches nor smiles; she mounts guard
on her convert as the night-long, soft
wham of the river-wind plucks at his eyelids
and rainwater rams
its endless meekness underneath the door.

from Punch's Gallery

Rouault: A Portrait

Outside by the sculpture garden, she tries to doze
in New York's grimy sun. To ease
the pain between her hips, she drags her knees
onto the iron bench, letting the breeze
lift her skirt. Perhaps she's guessed his ruse
is to watch from the window and feels his eyes
nudging her as he looks out, through a maze
of sundered spectra like a fool's disguise,
at her features clubbed with shadow. Booze
loosens them later; when he starts to seize
some advantage with blind craving, she goes
over to fill his glass. He shies,
but lets her pour. The whisky plays
rising notes in the bottleneck. A graze
from the metal slats she posed on, draws
his hand. The illness wells and flows,
bringing a fever with it, as she grows
whorish, eccentric, quick to please.

Ensor: Masks Confronting Death

A widow, her womb
shrunk to a walnut, hefts
her skirts, frolicking, wagging her hips
to seduce. She's rouged
like a whore. Will she put
some marrow back in those bones?

A girl in the flush of power,
impeccable, everything
watchable, lets her hand
be pressed to the naked teeth. She knows
she's not going home
with him tonight.

A murderer, the victim
still on his tongue,
looks at the eyes that look
through the girl's disguise. There follows
a brief, triangular
recognition. She shakes her head.

Punch, in the crowd with his stick,
is rabble-rousing,
as ever. His claque are baying; their sharp
snouts mark them out,
their whisky smell. He whacks
legs and grins invisibly.

Bonnard: Breakfast

1

An eyelid flickers, startled by the glass,
then droops in anticipation of the brush.

She starts on an indrawn breath. Her mouth goes slack.
Touch by touch she thickens every lash.

2

She does it naked. When she stoops to be
nose to nose with herself, her breasts make globes

clean of the rib-cage. There are poles of light
in the coppice beyond the lawn. A ground-mist ebbs.

3

Ovals and angles: he pictures the slope of her shoulder,
a hoisted hip, and lets a minute lapse

between placing their plates and knives and going to fetch
the basket of bread. A rhombus; a dark ellipse.

4

Water boils for the eggs. Chasing a thought
he sees it turn to confront him. *There's a smell* . . .

She is choosing a dress. He guesses she'll put it back
then choose another and put that back as well.

5

She might decide to stroll (he draws the bolt
on the tall french windows) shoeless across the grass

and then by the terrace, the stones holding her spoor.
She would see the table he's set: his place, her place.

6

She never looks in a mirror to brush her hair,
so she opens the blinds to view the day, hanging

her head to left and right for the volley of strokes.
The garden rocks on a pivot. Her ears are ringing.

7

It is whisky, the sourness rising in his nose
and scorching the membrane. His eye traces a smudge

above the skirting-board . . . She has come to the stair . . .
An ounce of glass litters the window-ledge.

8

He has posed her there, but can't make her descend,
or smile, or speak. He wonders, was it rash

to have picked that dress, to have left her feet unshod?
The welt on her cheek comes and goes like a blush.

Punch and the Gulls

One screamed from a chimney-pot, neck ducked,
spine taut, like a woman being fucked.

Punch howled and flapped his arms. She wouldn't fly.
Her world lapsed at the limit of her cry.

2

One harried a carcass on the beach,
hopped to and fro on tidemarked legs to reach

the riven torso, tore the flobby lung
and clapped the blood-clots underneath her tongue.

He watched another, frame by frame, displace
the arrested movements of a form in space.

She turned in the thermal, unperturbed, unjust,
lusted after, unimpeached by lust.

Punch as Victim

When she undressed, thoughtlessly, ready for bed,
taking her blouse off last, as always, her arse

a white pout topping the suntanned legs, he'd be there
with his imbecile grin; the little pool

of silk on the bedroom floor was his
to clean up. He'd pick hairs out of the bath

or open her bottles of cream when she wasn't there,
sniffing the surfaces, putting a dab

on the flaky skin above his jaw. One night
she bamboozled him with flattery; he slumped

onto the brace of her back in time to feel
the hiss of displeasure, a fleeting vibrato, fading

along the ripple of backbone. Her head stayed up;
she was looking straight at the wall, while he dropped

his cheek on her shoulder and stared at the bridge
her hand made on the sheet. The rings she wore

trickled light from stone to stone. The mash of his breath
was acrid, even to him. He unplugged and sat

back on his haunches as she subsided
and stayed there until she slept. Sometimes, he'd take

their photograph albums into his room and croon
with anticipation from moment to moment. The one

he liked best was a foggy shot—she was wearing his jacket
and scarf; if he half-closed his eyes

he could see himself, his fists dug into the pockets, the sly
courting of the camera, the moping lip. That day

the sun burned in the puddles. 'Remember,' he'd ask,
'the story I told on the phone?' She remembered it all

and watched him upend the decanter. His sleight of hand
was lifting terrible images from a glance

to left or right, from an absence, from some coy
pose as she window-shopped. His memories

ran with him like a dog-pack. The glorious prospect of sin
and discovery kept him alert. He'd lie

close as a poacher to catch her
at the sink, side-lit, applying a *gant de toilette*,

knees splayed like a dancer's, her features set
in vacuity, seamless, unransomed by hope or regret.

Punch's Nightmares 2

He was a mouth on a pole
shouting orders from the dug-out.
The mud looked like melted coffee ice-cream.

When the noise stopped
he thought about getting back
to wherever he'd been when he left her.

The officers were dead.
If she'd really known them all, then why,
he wondered, wasn't she here

to keen over the rumpled gas-bags of bodies?
He stood in a nest of sinew
soppy with blood

and sniffed the air.
A pace or so from the trench, she stopped
and watched the poison billow in the craters,

hearing him say, again, that he'd never desert.
He framed the distinguished face
between two ruptured tree trunks, just in case

he'd have to remember her like that.
'You're okay,' she said. 'What a place!' A cadaver
stirred in the mire.

Punch's Nightmares 3

First it was a railway station. He admired
the frogging on his father's uniform, noted the way
the old man tapped the fringe of his moustache

with the topside of a finger when they'd kissed.
The train pulled out. He was surprised
to find himself the passenger.

A dark-room next.
He stood behind the hands of the technician
and watched the damson-coloured mark

grow on her back like a flaw in the paper.
The photograph turned sepia; a text
and caption fixed the place and used his name.

Then a fashionable street. The mannequins,
with perfect nipples and nails,
fixed blind eyes on the spotlights . . .

Behind plate-glass, surrounded
by the soft hues of the season, Punch
poured tea for the Femmes de Venise.

His father stopped to peer in, cracking
a boot with his swagger-stick. The women
swivelled their heads like birds.

She offered a scone and jam. The livid pulp
of burn-scars, the tunnel-nostrils, gleamed
like tin-foil. He took the food and ate.

Punch's Nightmares 4

She sloughed the old skin off.
The new skin glistened, transparent:
a smell of paraffin.

Between the purple coils
and bulbous tucks of entrail
he could see his children backed up

in her womb. She lay
on the dampened sheet, gin-clear,
a weft of muscle

pursing to make a frill
that ran on her belly, like
an undersea current.

The mesh of hair in her groin,
the blots of her nipples,
were delicate abstracts. She'd lost

years of soil: handclasps,
the courtesies of dancing partners,
hapless gestures, the tincture

flesh leaves on flesh.
The straps and ribbing of clothes
would have bruised her; perfume

would have smoked on her wrists and neck.
From the pearl of her heel to the soft
fingernails, she was untouched.

He dreamt he was dreaming all this
and waited to wake
from his own weightless tread,

from the blue stage-glow
that lit the dream, from sound
suspended, until he saw

her lips, bonded
like blown glass, incorruptible, and her eyes
white with ecstasy.

Punch's Nightmares 7

The blind backs of houses
formed endless alleys, smooth
and pitted as ancient leather.
There were soldiers and flares.
He loped past with his hound.

She entered to applause,
courting the footlights, following
a cart. She was sifting
the faces of the near-dead.

He walked in the wheel-ruts
to where they'd stopped for the night.
He peered over
the stable door. The driver
was at her like a boar.

Next morning she rode
on the tailgate, trailing her legs.
The bodies clattered like laths.
He saw them off
into the wings. She'd never
find him among the stiffs.

Now he came centre-stage.
Now his soliloquy,
his grace-note, the moment of moments
when he'd gloss the conceit.
The packed house hung
on his indrawn breath. Backstage
the soldiers ran
from door to door with their torches.

The Blessed Punchinello, Mart.

He would live in unapproachable light
and the numbing silence of denial,
weathered in desert air
to leather and bone, a shrivelled sun dial

welded by bone to the pillar: and record
rumours of the world
as a feather on a stripped nerve.
His hands, both curled

like tubers in his lap,
would cease to learn
the wintry rituals of supplication.
Every day, his lean

shadow would run to lodge its tip
like a splinter beneath her skin.
He'd grow remote
and unignorable, the thin

end of a wedge, a silenced voice
listened-for like the plummet of an echo
when the land-line shrinks,
when the city starts to glow

raising a false dawn and the hum
of vast machines
puckers the skin on the river.
He would know what reduction means,

the beauty of anorexia, the slow
smelting of martyrs! A fever
would scorch his eyes, the noon sky
flare with visions, then shrink to the shiver

and rustle of gnats at evening.
The scratch of his stubble growing, the least
tremor on the air, would touch his senses:
the lightest breath of the beast

splashing the plinth below
then trotting towards the suburbs, her spoor
hot in its nostrils, a dream of extravagance
swamping the pink-black mottle of its jaw.

Mr Punch Confronts his God

The stonework soared,
but inch by bloodstained inch. His shout
churned in a vacuum of arches and rafters.

A pain welled in his shoulder and ran
the length of the bones in his arm;
he flexed his fist to let it find a level;

it sank to a low ache,
tugging his thumb in a spasm that matched
his heartbeat. He felt

as if he'd been tranced. His voice
and the sharp flow in his arm
seemed an awakening.

Deep in the swimming particles of dusk
the dying figure
throbbed on the cross. Who could survive

that ardour? Not popes, not kings,
not Cuthbert on his rock
flayed by the northern winds,

pinned there by troth, perceiving
an awesome response in the ocean rising
and running bleak green on the stone . . .

A wind lifted the spray: a distant whisper
no louder than bridehood closing
its petals to seal a canker; then

all fear, all morbid love, all words
to celebrate, all paint to thumb the eye,
all music's power to stun,

rose in his mind, gathered, and was lost:
a knowledge he'd never owned
nor sought. A light came on

in the porch, so he walked to the door,
his heeltaps ringing like asdic in the flags.
A glow from the rich glass patched

the dark facade. In every cleft
angels clung, or swarmed on the wall like bats.
Some sulphurous eye was watching as he left.

Punch at Peace with Himself

The particles slowed
all in harmony. The weft
of hair on his forearm gleamed
like wheat turned by a wind.

His apeish thoughts went
out of focus
then ran together, a soup
of tints and tones.

He hummed one note:
a dynamo, shutting down.
He lay in lines. His limp
dick was a damp squib.

His lungs puckered, just.
The piebald, waxy rill
on his jawline and throat
was fats and fluids setting.

Things gathered to leave.
The garden was shut in
by rain; he stared
at the water-lights and grinned

a baby's grin. Keeping discreet
station behind his eye-line
(her skirts rustled) sat
the beloved executrix.

Punch's Day Book

'They've stacked your pictures in among the books.
There are flowers and cushions, but nobody goes in
to work or read. The house
is full of music always.
                              Will you become
a monk, with your manuscripts
and your winnowing, skin-deep lust?
Will you never
recover from your life? At every turn
there is something
to tear you, or make you afraid.'

*

'Almost nothing moves
in your universe of phantoms. Nothing revives.
A lifetime's words have puddled in you, like gall.
Not fury, but the residue of fury.
Not love, but love's cold junk of bones.
                              Pretend
the window has been flung back; you are standing
close to the downpour. Everything floods with green
There's no beginning
and no end to the moment.'

*

'Mornings are worse. Do you
find that? I wake as light arrives
at every crevice in the house. The birds
are a chorus of women
thrilled by new possessions. I might hear
the church bell, an aeroplane
descending in steps ... The world
is close, but outside:
beyond the window-sill, beyond the gate,
a guileless cryptogram.

I shall die of my thoughts. I've become
my story's heroine
saturated by disease, the last
of the beautiful tuberculars.

All this would be solved
if I had you between the pulses of my wrists.'

*

'When you pray, what do you want?
I ask the almost-possible.
My days are carried on the backs of birds.
I can feel a purpose about me, undefined,
some mornings, on waking . . . some mornings, not.
It is folly to doze till noon and folly not to.
I lie in a pouch of odours, planning risks
and the world saunters past. Last week
I dreamt we were young, the children were young,
it was autumn, an evening, heat still lay
in the walls of our house,
the children slept on our laps,
damp, white curls, the tiny stitching of breath,
and we cradled our first drinks of the day . . . Of course,
I dreamt no such thing, not I.
The dogs go everywhere with me now.
I return from my morning ride
business-like with parings of hoof and horn
for the compost's mulch. I'm a Victorian lady
working my garden, wearing
broderie-anglaise and my white straw hat,
leaving slipper marks in the dew.
If you could see me
wandering here! I pause
on the bridge over the stream, or settle
with book and glass
beneath the pear tree's quaint strappado.'

*

The hang of her brow
over a book, her wrists
crossed one on the other below the page,
or the vent of an eye
as she looked away, revealed
the bloodline of Saxonish neurotics.

There was low music, slow,
caught in the roll of woodsmoke
blotting the summer garden.
Something made her look away,
sharply, towards the stream
and the circle of chairs beside the orchard.

One moment reading, wrists loosely crossed,
the next, head lifted, alert,
the pose of a creature
disturbed at its feeding: a fixed
stare; then, reassured, she dropped her gaze
to the page, half-hearing the music perhaps.

You could see the blush of health
beneath the furze on her cheek,
the brightness of the creature's eye.

'That summer, our youth astounded me.
The house still fixes my dreams; I read
or walk by the stream, past the orchard.

You were edgy, or so it seemed;
eager to be away.
Then, as now, your words impoverished me.'

*

'I hoard my luck: a brooch, a knot of straw. I try
to marry the tune to the rhyme. Sometimes
I move things from here to there—
a housewife's coup d'etat.
                                    The skipping-song
had a beat like a muffled drum. You brought
food on a tray. The children released
their starry, gluttonous smiles into my care.

The writing-desk shall be so,
this chair might stand
beside the hearth, the pictures shall be so,
and each room have its bowl of pot-pourri.'

*

'I turn the page. Smoke streams in my lungs
leaving them damp and stretched. All human shock
gathers in the chill of three a.m. Across the county
shutters are closed against the wind. Some menace,
lean with spite, comes with the early silence
as if wolves still ran the hillsides, quartering
an old killing ground, the skyline shrunk
to cameo inside a yellow iris.'

*

'My lover has foxy teeth and keeps two dogs.
Perhaps you remember him.
He'll hoist me over a cask and lift my skirt
at any time.
                    It all goes well,
you can see I'm feeling well.
The dogs hunt every spinney if I slip them
and come back flecked with blood.
They said the weather would break.
                                        I won't have children
whatever tricks he tries. To be outlived
is terrifying.
                    At times
the summer seemed to have lasted half the year.
Really, I came to think
I could be here forever, the valley's stoic
gathering small virtues, a student of leaf and fruit.
The ascetic wants to shine in the eye of God.'

*

'I try to progress my life minute by minute.
The striking clocks have been sold, so I cannot tell
how long it takes you
to do the things I invent for you to do.
The people I give you to meet
are witty and anchored to circumstance.

I shall make a nursery rhyme
about the creature in this Christmas photograph:
mop-haired, the scourge of children.
My dreams deliver confidences.
I am ransacked, I am open to the weather,
I echo with pleasantries.

My next task is to sort
the dead images into piles—with good grace.
When you read this, picture me
serene beneath my sheet,
at my foot, the cocktail-shaker; at my head
a table of oils, guaranteeing preservation.'

*

'There are those who plan to die
blameless, open-handed, an unwritten letter.
We can't aspire to that.
We lack the pure compulsion and the nerve.

The orchard's harvested; the stoves are lit
to burn all winter; the house is steeped
in a musty odour of fruit.
                    Think how it is
to own nothing, to carry nothing
from one place to the next . . .
Unburdened, my body grows
featureless. I could disappear in water,
be perfectly matched to grassland.
                    Every tree
is stripped and life goes on underground;
even the telephone's in hibernation.

I shall be here, of course,
seeing the season out from my fireside chair,
sometimes bringing apples down from the loft
or walking to church. If I should stray,
how would you ever find me?—
a pallid silhouette
on a clear road, like any refugee.'

*

'A night-long blow, seeded with sleet
as if a door had been opened on the north . . .

I am mobbed by shadows.
The future is codified

a serial-number die-stamped on my bones.'

*

'The Devon love-knot
the gleaming bevel
of the ha-ha, the beeches, the pic of you
emerging with a tray, ha-ha

the pic, the snap
the children dancing
your finger fingering my quim, ha-ha
ha-ha, the knot, the bright bevel

ha-ha'

NEW POEMS

The Windhound

1

Even from here
the sea is a magnet.

Her shadow
falls on my shadow; she leans
towards me like a confidante
with news of her foragings
and the rowan's scandalous fruiting.

Month by month she's attended me,
in her mouth some small dead thing
dishevelled from the course.

The season frets in its traces.
What will the sea have
but its share of the year's cull.

2

Early dusk, and her blood's up
for the scents and sudden movements.

This secret life, her dark route,
is the nub of her parentage.

It's all of a piece:
lofted muscle, cranked nerve, a sprint uphill,

the headiness of the first drink—
gin over ice—her gum

shrinking at the sharp
silken taste of it.

3

E-flat D-natural
will always fetch her;
it could be anyone
working her in the half-dark.

She goes head down
sampling hot dropings,
the epicure's
fastidious purse and nibble, ,

then gets to the crest,
wide-eyed, in a single rush,
passion's tinnitus piping
at either temple.

The risk is in losing touch
before she's ready
to come by with her kill,
but I gauge the line she'll take

and from time to time
we gain a sense of each other—
the amiable butcher,
the beacon's profiteer.

4

She fell on the creature's neck and bit
just behind the ear ... there ... Her whisper
was intimate and orthodox. She squatted
and set her tongue to drag against the nap.

Now it's hulked. I place the heart and liver
on my flattened palm. She nips them from the gore
and smothers her palate. A waisted flute
of cold champagne is required to rinse her teeth.

Soon she'll go to the couch and start to groom
the tapering russet and black of her handspan loins,
the pubic teasel, then turn to face the fire
and dream of herself and me in different guises.

5

A change in the weather
will madden the birds. This one
dips into its plumage,
unhinged by the crimson lights behind its eyes.

Her valley holds
the druggy odour of an apple-store;
fog blooms on the watercourse;
a smoking blackthorn
is the hedgerow's Hand of Glory.

Narcolepsy pins her to the mat.
A slut's debris
litters the floor:
a meal half-eaten, a brush and comb, a letter
dropped by the grate.
. . . *nothing remains untouched.*

She could only quicken, now,
to thoughts of the city,
her nails ticking on parquet,
intoxicating furs in the arcades.

6

The hillside was a tenement.
She used to eavesdrop
on muffled subterranean lives.

Everything in that place
judged its scale
and growth by her silhouette.

She had sifted the landscape
for what was usable.
Her private life

was morse in the furrows, a grid
of sightlines
and the rough nudge of instinct.

To reconstruct such loss
image by image
is the work of a lifetime.

Think of it: you must start
with a heat-haze
from the valley's forge,

the boss of her hip
smooth
and canted like an anvil.

7

There she is, absorbed
by a wall of mirrors,
the party's mascot
in love with whoever feeds her.

What will the gossip be
of this lean adventuress
turning rump to rump
with her likeness?

The room is a dish of light.
She goes from hand to hand,
a rind of canape
stuck to her teeth like smegma.

Now and then she startles herself:
tipped to the edge of a run
by the soft white scut
of a dropped glove,

or lifting her long nose
to a thin stink of moss,
tart as iodine,
from the air-conditioning;

but she battles that.
She wants to be a blueprint
for the well-bred, a skin-flick
for the pot-valiant.

They all sidle round in a ring.
Their smiles
could grind glass.
Their hands ravel goodwill.

She lopes home,
headachey, stagestruck,
and goes to the bed
and lies as if she'd been hurled there.

If you bent your head to her head
while she slept, you'd trace the faint
corrosive heat
of nicotine on her breath.

8

Look down from the ridge. The valley's frost-hollows
are the eye-sockets of a vast chrysalis
wrapped in the acreage, its pulse close to zero.

Translate that meadow's bald archaeology—
the trace of some swiftness that flowed downhill,
ley-lines of the hare's queer pilgrimage.

Absentmindedness might bring you here.
The heron's wingbeats are cold lacunae.
The copse is a dropped portcullis.

This stone is just so.

9

(*1*)

The bar-room cirrus is volatile.
She sifts the odours: Scotch,
jet-fuel, and something hot
like camphor. The young execs
are devilish with drink. Her fantasy
absorbs the moon-faced clock
and the blaze of the barman's strawberry-mark.

93

'Only did three lines—Christ,
I was st*oo*pified!
Where d'you *find* that stuff?'

She had surrendered her name
to the quick rich,
elfin at cocktail hour,
her face drubbed by a slipstream
as the counties clicked past.

The world indoors became
rash with hearsay: a sly patois,
the bloodless wager of the orgiast.

'Same again My Pet . . . Dear God,
you should have seen
the tits on it . . . arse up,
head down, and never gave a yelp.
Look, make
those doubles, will you . . . ?

'I know, I know;
I've heard all that before.'

(2)

She goes by wharves and car dumps,
moonblind, by movie-queues,
by sweatshops and peepshows,
nosing the pale yolks
of sputum on the kerbstones.

Here are the migrants
flocking to the light,
restored by new friendships.

Here are the boys and girls
whey-faced with giving.

94

Here are the bag ladies
and the old men with bottles,
roused by the sudden warmth
as the city switches on,
back-alley hibernators, bringing
their little flogged faces
to the burrow's edge.

Poaching their precinct
she can flush
from a stew of garbage,
the rat and the rat's get.

(*1*)

She flukes a dream:
the weir a centrifuge
to the luminous green
heft of water
stilled by repetition.

Her magnified eye
shuttles across the bright
warp of the river weed.
She has gone
underneath the pour.

A convert's snapshot vision
clings to the meniscus—
tree canopy
and cloudbank
washed off the iris.

There is nothing here
to own, except
the mechanics of immersion,
the scrupulous laws
of the prism.

(*2*)

The valley-side is backed up to the moor,
the moor to the sea, the sea to rough weather.

What ripened here is a stain gone to the roots.
A seismic percussion of seeding underground

registers in the long bones of her feet
and the dream darkens as a squall

rushes the slope. She walks in the piling wind
half-visible, half-conscious, the terrain

mapped by her sleeping eye, and goes between
the dripping lines of contour like a haunting.

(*3*)

She crosses the blunt stubble
and pulls up,
amazed by her own arrival.

A buzzard is circling
under the sun, its broad
pinion-feathers flaming,

cattle are standing
head to head in the low pasture,
fixed by the heat—

everything changed,
everything just as she left it.
I can almost see her

stepping out of shade
beside the copse
head bowed, like a woman

hunting the parched grass
for the glint
of a dropped earring.

This dream is all her luck
caught in the balance:
a chorus from the baked earth,

something that stirs in her gaze
beneath the hill's brim,
whatever is cold on the tongue.

The Analysand

Just on the cusp of sleep
the image of a hare, hunkered
in the lee of a blackthorn hedge,

a sloping snowfield, a spinney,
the moon like a crooked sixpence ...
She'd expected to know the place.

It was dawn from the smell
of bacon in the pan
and the brisk riddling of coals.

She fetched him out, his boots
breaking the snowcrust
in Church Lane, the dogs at heel;

and, oh, he did it perfectly—
clapping against the cold
so that the sound

could reach her at one remove;
stopping to watch a heron;
lifting one hand, like someone

bowling under-arm,
to release the dogs. A rustle
of breath like silver leaf

touched his lips as he started
to tackle the hill.
In that air, she could see

prisms in the spindrift
off his toecaps. No matter;
she was smug with speed.

As he stamped up
the steepest pitch, just past
the spinney and rose

towards her (so close
that she almost laughed)
she leapt into wakefulness.

*

I can't tell why,
but the most important part
was fetching him out:

his handclap coming at me
a pulse-beat late, the way
he set the dogs running.

I was lying up
in the lee of a hedge; even so,
I could see everything

as if I sat on his shoulder;
and it came, remember,
on the cusp of sleep.—

Didn't you say
those are the truest dreams?
Well, I was puss,

a flibbertigibbet, familiar
to some wise old woman.
What do you make of it? Is he still

working the dogs on the down?
Will my children be harelipped
and my gaze mildew the grain?

Playback

On the crown of the moor
the wolf takes
his own axis forward
again and again and runs
through it again and again

in bad light,
in a drizzle of static, his eyes
snapping on and off.
He is the last wolf, and these
are the final frames—

the wolf, the moor, the gun, the spoor
and man with his two hands.

*

Or there are neat
compositions: just sky
and the hawk's head, motionless
while he treadles the updraught, a perfect
target on looped tape

and the hare, fixed
in mid-gallop
on the borrow of the hill,
her long paddles airborne; then
we get the group shot—

hawk and hare, gibbet and snare
and man with his two hands.

*

Now selected scenes
from a very long story. Fields
under the plough; wet furrows
stripping silver rods
across the eyeline;

the seed rooting; the shoot
unpicking; cloud after cloud
belly-down; and rain
hammering the drills.
Here is a brief synopsis—

the land, the root, the poisoned shoot
and man with his two hands.

2

This is the place at one remove
where you watch the sea fill up and smudge
like an infection flooding

with plum-dark blood ... You can kick
your heels by the ash-pits.
You can lob stones at the slow

cloacal yawn of the breakers.
You can ask yourself
how in the world you got here

and whether you'll ever get back.
Everything you see
has abandoned everything else.

The birds on the foreshore
are mad ascetics. Simply feather and bone,
they traipse the sand-flats

mistaking their food for money.
There's a silence
like the silence after a crescendo

or a bad accident; and your voice
if you speak
is the strangest thing of all.

3

This is a version
run for children and simpletons.
The wolf backtracked
in a celluloid blizzard, picking
his pawprints off the snow;

the hawk decoded the tape;
the hare tottered off
like a joker, her grin
stuck to the lens; the clouds
were a fata morgana.

Man dropped his hands.
All the hospitals emptied
and philosophy spawned in the sea.

OXFORD POETS

Fleur Adcock

James Berry

Edward Kamau Brathwaite

Joseph Brodsky

Michael Donaghy

D. J. Enright

Roy Fisher

David Gascoyne

David Harsent

Anthony Hecht

Zbigniew Herbert

Thomas Kinsella

Brad Leithauser

Herbert Lomas

Derek Mahon

Medbh McGuckian

James Merrill

John Montague

Peter Porter

Craig Raine

Tom Rawling

Christopher Reid

Stephen Romer

Carole Satyamurti

Peter Scupham

Penelope Shuttle

Louis Simpson

Anne Stevenson

George Szirtes

Anthony Thwaite

Charles Tomlinson

Chris Wallace-Crabbe

Hugo Williams

also

Basil Bunting

W. H. Davies

Keith Douglas

Ivor Gurney

Edward Thomas